THE INDEPENDENT SET ALGORITHM

ASHAY DHARWADKER

ABSTRACT

We present a new polynomial-time algorithm for finding maximal independent sets in graphs. It is shown that every graph with n vertices and maximum vertex degree Δ must have a maximum independent set of size at least $\lceil n/(\Delta+1) \rceil$ and that this condition is the best possible in terms of n and Δ. As a corollary, we obtain new bounds on the famous Ramsey numbers in terms of the maximum and minimum vertex degrees of the corresponding Ramsey graphs. The algorithm finds a maximum independent set in all known examples of graphs. In view of the importance of the **P** versus **NP** question, we ask if there exists a graph for which the algorithm cannot find a maximum independent set. The algorithm is demonstrated by finding maximum independent sets for several famous graphs, including two large benchmark graphs with hidden maximum independent sets. We implement the algorithm in C++ and provide a demonstration program for Microsoft Windows.

The Demonstration Program

http://www.dharwadker.org/independent_set

CONTENTS

1. Introduction

In 1972, Karp [1] introduced a list of twenty-one **NP**-complete problems, one of which was the problem of finding a maximum independent set in a graph. Given a graph, one must find a largest set of vertices such that no two vertices in the set are connected by an edge. Such a set of vertices is called a maximum independent set of the graph and in general can be very difficult to find. For example, try to find a maximum independent set with five vertices in the Frucht graph [2] shown below in Figure 1.1.

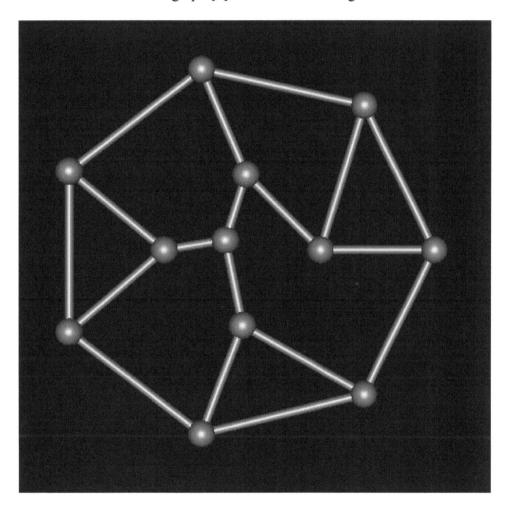

Figure 1.1. *Find an independent set with five vertices*

We present a new polynomial-time **INDEPENDENT SET ALGORITHM** for finding maximal independent sets in graphs. In Section 2, we provide precise **DEFINITIONS** of all the terminology used. In Section 3, we present a formal description of the **ALGORITHM** followed by a small example to show how the algorithm works step-by-step. In Section 4, we show that the algorithm has polynomial-time **COMPLEXITY**. In Section 5, we give a new condition of **SUFFICIENCY** for a graph to have a maximum independent set of a certain size. We prove that every graph with n vertices and maximum vertex degree Δ must have a maximum independent set of size at least $\lceil n/(\Delta+1) \rceil$ and that the algorithm will always find an independent set of at least this size. Furthermore, we prove that this

7

condition is the best possible in terms of n and Δ by explicitly constructing graphs for which the size of a maximum independent set is exactly $\lceil n/(\Delta+1) \rceil$. As a corollary, we obtain new bounds on the famous Ramsey numbers in terms of the maximum and minimum vertex degrees of the corresponding Ramsey graphs. For all known examples of graphs, the algorithm finds a maximum independent set. In view of the importance of the **P** versus **NP** question **[3]**, we ask: *does there exist a graph for which this algorithm cannot find a maximum independent set?* In Section 6, we provide an **IMPLEMENTATION** of the algorithm as a C++ program, together with demonstration software for Microsoft Windows. In Section 7, we demonstrate the algorithm by finding maximum independent sets for several **EXAMPLES** of famous graphs, including two large benchmark graphs with hidden maximum independent sets. In Section 8, we list the **REFERENCES**.

2. Definitions

We begin with precise definitions of all the terminology and notation used in this presentation, following **[4]**. We use the usual notation $\lfloor x \rfloor$ to denote the *floor function* i.e. the greatest integer not greater than x and $\lceil x \rceil$ to denote the *ceiling function* i.e. the least integer not less than x.

A *simple graph G* with n vertices consists of a set of *vertices V*, with $|V| = n$, and a set of *edges E*, such that each edge is an unordered pair of distinct vertices. Note that the definition of G explicitly forbids *loops* (edges joining a vertex to itself) and *multiple edges* (many edges joining a pair of vertices), whence the set E must also be finite. We may *label* the vertices of G with the integers 1, 2, …, n. If the unordered pair of vertices $\{u, v\}$ is an edge in G, we say that u is a *neighbor* of v and write $uv \in E$. Neighborhood is clearly a symmetric relationship: $uv \in E$ if and only if $vu \in E$. The *degree* of a vertex v, denoted by $d(v)$, is the number of neighbors of v. The *maximum degree* over all vertices of G is denoted by Δ. The *adjacency matrix* of G is an $n \times n$ matrix with the entry in row u and column v equal to 1 if $uv \in E$ and equal to 0 otherwise. A *clique Q* of G is a set of vertices such that every unordered pair of vertices in Q is an edge. A *vertex cover C* of G is a set of vertices such that for every edge $\{u,v\}$ of G at least one of u or v is in C. An *independent set S* of G is a set of vertices such that no unordered pair of vertices in S is an edge. Given an independent set S of G and a vertex v outside S, we say that v is *adjoinable* if the set $S \cup \{v\}$ is still an independent set of G. Denote by $\rho(S)$ the *number of adjoinable vertices* of an independent set S of G. A *maximal independent set* has no adjoinable vertices. A *maximum independent set* is an independent set with the largest number of vertices. Note that a maximum independent set is always maximal but not necessarily vice versa.

An *algorithm* is a problem-solving method suitable for implementation as a computer program. While designing algorithms we are typically faced with a number of different approaches. For small problems, it hardly matters which approach we use, as long as it is one that solves the problem correctly. However, there are many problems for which the only known algorithms take so long to compute the solution that they are practically useless. A *polynomial-time algorithm* is one whose number of computational

steps is always bounded by a polynomial function of the size of the input. Thus, a polynomial-time algorithm is one that is actually useful in practice. The class of all such problems that have polynomial-time algorithms is denoted by **P**. For some problems, there are no known polynomial-time algorithms but these problems do have *nondeterministic polynomial-time algorithms*: try all candidates for solutions simultaneously and for each given candidate, verify whether it is a correct solution in polynomial-time. The class of all such problems is denoted by **NP**. Clearly **P** $\subseteq$ **NP**. On the other hand, there are problems that are known to be in **NP** and are such that any polynomial-time algorithm for them can be transformed (in polynomial-time) into a polynomial-time algorithm for every problem in **NP**. Such problems are called **NP-**
complete. The problem of finding a maximum independent set is known to be **NP-**
complete [1]. Thus, if we are able to show the existence of a polynomial-time algorithm that finds a maximum independent set in any graph, we could prove that **P** = **NP**. The present algorithm is, so far as we know, a promising candidate for the task. One of the greatest unresolved problems in mathematics and computer science today is whether **P** = **NP** or **P** $\neq$ **NP** [3].

3. Algorithm

We now present a formal description of the algorithm. This is followed by a small example illustrating the steps of the algorithm. We start by defining two procedures.

3.1. Procedure. Given a simple graph G with n vertices and an independent set S of G, if S has no adjoinable vertices, output S. Else, for each adjoinable vertex v of S, find the number $\rho(S \cup \{v\})$ of adjoinable vertices of the independent set $S \cup \{v\}$. Let v_{max} denote an adjoinable vertex such that $\rho(S \cup \{v_{max}\})$ is a maximum and obtain the independent set $S \cup \{v_{max}\}$. Repeat until the independent set has no adjoinable vertices.

3.2. Procedure. Given a simple graph G with n vertices and a maximal independent set S of G, if there is no vertex v outside S such that v has exactly one neighbor w inside S, output S. Else, find a vertex v outside S such that v has exactly one neighbor w inside S. Define $S^{v,w}$ by adjoining v to S and removing w from S. Perform procedure 3.1 on $S^{v,w}$ and output the resulting independent set.

3.3. Algorithm. Given as input a simple graph G with n vertices labeled 1, 2, …, n, search for an independent set of size at least k. At each stage, if the independent set obtained has size at least k, then stop.

- **Part I.** For $i = 1, 2, ..., n$ in turn
  - Initialize the independent set $S_i = \{i\}$.
  - Perform procedure 3.1 on S_i.
  - For $r = 1, 2, ..., k$ perform procedure 3.2 repeated r times.
  - The result is a maximal independent set S_i.
- **Part II.** For each pair of maximal independent sets S_i, S_j found in Part I

- Initialize the independent set $S_{i,j} = S_i \cap S_j$.
- Perform procedure 3.1 on $S_{i,j}$.
- For $r = 1, 2, ..., k$ perform procedure 3.2 repeated r times.
- The result is a maximal independent set $S_{i,j}$.

3.4. Example. We demonstrate the steps of the algorithm with a small example. The input is the Frucht graph **[2]** shown below with $n = 12$ vertices labled

$$V = \{1, 2, 3, 4, 5, 6, 7, 8, 9, 10, 11, 12\}.$$

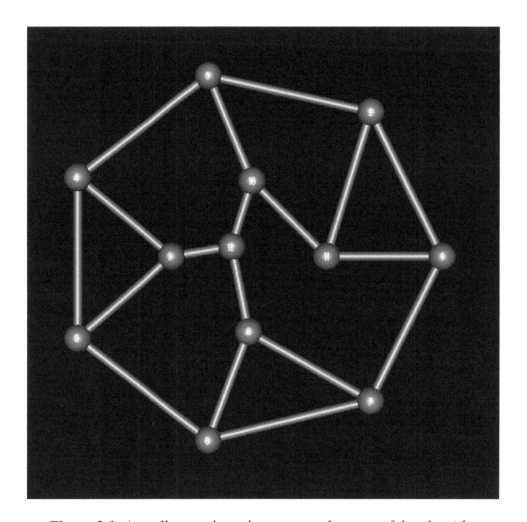

Figure 3.1. A small example to demonstrate the steps of the algorithm

We search for an independent set of size at least $k = 5$. Part I for $i = 1$ and $i = 2$ yields independent sets S_1 and S_2 of size 4, so we give the details starting from $i = 3$. We initialize the independent set as

$$S_3 = \{i\} = \{3\}.$$

We now perform procedure 3.1. Here are the results in tabular form:

Independent Set $S_3 = \{3\}$. Size: 1.

Adjoinable vertex v of S_3	Adjoinable vertices of $S_3 \cup \{v\}$	$\rho(S_3 \cup \{v\})$
1	5, 6, 8, 9, 12	5
5	1, 7, 8, 11, 12	5
6	1, 9, 11, 12	4
7	5, 9, 11, 12	4
8	1, 5, 9, 11	4
9	1, 6, 7, 8, 11	5
11	5, 6, 7, 8, 9, 12	6
12	1, 5, 6, 7, 11	5

Maximum $\rho(S_3 \cup \{v\}) = 6$ for $v = 11$. Adjoin vertex 11 to S_3.

Independent Set $S_3 = \{3, 11\}$. Size: 2.

Adjoinable vertex v of S_3	Adjoinable vertices of $S_3 \cup \{v\}$	$\rho(S_3 \cup \{v\})$
5	7, 8, 12	3
6	9, 12	2
7	5, 9, 12	3
8	5, 9	2
9	6, 7, 8	3
12	5, 6, 7	3

Maximum $\rho(S_3 \cup \{v\}) = 3$ for $v = 5$. Adjoin vertex 5 to S_3.

Independent Set $S_3 = \{3, 5, 11\}$. Size: 3.

Adjoinable vertex v of S_3	Adjoinable vertices of $S_3 \cup \{v\}$	$\rho(S_3 \cup \{v\})$
7	12	1
8	None	0
12	7	1

Maximum $\rho(S_3 \cup \{v\}) = 1$ for $v = 7$. Adjoin vertex 7 to S_3.

Independent Set $S_3 = \{3, 5, 7, 11\}$. Size: 4.

Adjoinable vertex v of S_3	Adjoinable vertices of $S_3 \cup \{v\}$	$\rho(S_3 \cup \{v\})$
12	None	0

Maximum $\rho(S_3 \cup \{v\}) = 0$ for $v = 12$. Adjoin vertex 12 to S_3.

We obtain a maximal independent set

$$S_3 = \{3, 5, 7, 11, 12\}$$

of the requested size $k = 5$ and the algorithm terminates.

4. Complexity

We shall now show that the algorithm terminates in polynomial-time, by specifying a polynomial of the number of vertices n of the input graph, that is an upper bound on the total number of computational steps performed by the algorithm. Note that we consider

- checking whether a given pair of vertices is connected by an edge in G, and
- comparing whether a given integer is less than another given integer

to be *elementary computational steps*.

4.1. Proposition. Given a simple graph G with n vertices and an independent set S, procedure 3.1 takes at most n^5 steps.

Proof. Checking whether a particular vertex is adjoinable takes at most n^2 steps, since the vertex has less than n neighbors and for each neighbor it takes less than n steps to check whether it is outside the independent set. For a particular independent set, finding the number ρ of adjoinable vertices takes at most $n^3 = nn^2$ steps, since for each of the at most n vertices outside the independent set we must check whether it is adjoinable or not. For a particular independent set, finding a vertex for which ρ is maximum then takes at most $n^4 = nn^3$ steps, since there are at most n vertices outside. Procedure 3.1 terminates when at most n vertices are adjoined, so it takes a total of at most $n^5 = nn^4$ steps. $\square$

4.2. Proposition. Given a simple graph G with n vertices and a maximal independent set S, procedure 3.2 takes at most $n^5 + n^2 + 1$ steps.

Proof. To find a vertex v outside S that has exactly one neighbor w inside S takes at most n^2 steps, since there are less than n vertices outside S and we must find out if at least one of the less than n neighbors of any such vertex are inside S. If such a vertex v has been found, it takes one step to exchange v and w. Thereafter, by proposition 4.1, it takes at most n^5 steps to perform procedure 3.1 on the resulting independent set. Thus, procedure 3.2 takes at most $n^2 + 1 + n^5$ steps. $\square$

4.3. Proposition. Given a simple graph G with n vertices, part I of the algorithm takes at most $n^7 + n^6 + n^4 + n^2$ steps.

Proof. At each turn, procedure 3.1 takes at most n^5 steps by proposition 4.1. Then procedure 3.2 is performed at most n times, since k can be at most n. This, by proposition 4.2, takes at most $n(n^5 + n^2 + 1) = n^6 + n^3 + n$ steps. So, at each turn, at most $n^5 + n^6 + n^3 + n$ steps are executed. There are n turns for $i = 1, 2, ..., n$, so part I performs a total of at most $n(n^5 + n^6 + n^3 + n) = n^6 + n^7 + n^4 + n^2$ steps. $\square$

4.4. Proposition. Given a simple graph G with n vertices, the algorithm takes less than $n^8 + 2n^7 + n^6 + n^5 + n^4 + n^3 + n^2$ steps to terminate.

Proof. There are less than n^2 distinct pairs of maximal independent sets found by part I, that are treated in turn. Similar to the proof of proposition 4.3, part II takes less than $n^2(n^5 + n^6 + n^3 + n) = n^7 + n^8 + n^5 + n^3$. Hence, part I and part II together take less than a grand total of $(n^7 + n^6 + n^4 + n^2) + (n^8 + n^7 + n^5 + n^3) = n^8 + 2n^7 + n^6 + n^5 + n^4 + n^3 + n^2$ steps to terminate. $\square$

4.5. Remark. These are pessimistic upper bounds for the worst possible cases. The actual number of steps taken by the algorithm to terminate will depend on both n and k. For smaller values of k, the algorithm terminates much faster. In almost all of the examples in section 7, one or two steps of part I already find a maximum independent set. Only the second benchmark, Witzel's graph 7.20, requires part II of the algorithm to find a maximum independent set.

5. Sufficiency

The algorithm may be applied to any simple graph and will always terminate in polynomial-time, finding many maximal independent sets. The propositions below establish sufficient conditions on the input graph which guarantee that the algorithm will find maximal independent sets of a certain size. Specifically, we prove that every graph with n vertices and maximum vertex degree Δ must have a maximum independent set of size at least $\lceil n/(\Delta+1) \rceil$ and that the algorithm will always find an independent set of at least this size. Furthermore, we prove that this condition is the best possible in terms of n and Δ by explicitly constructing graphs for which the size of a maximum independent set is exactly $\lceil n/(\Delta+1) \rceil$. As a corollary, we obtain new bounds on the famous Ramsey numbers in terms of the maximum and minimum vertex degrees of the corresponding Ramsey graphs. The proofs use two fundamental axioms: Euclid's Division Lemma [5] and the Pigeonhole Principle [6].

Euclid's Division Lemma. Given a positive integer m and any integer n, there exist unique integers q and r with $0 \le r < m$ such that $n = qm+r$.

Pigeonhole Principle. If l letters are distributed into p pigeonholes, then some pigeonhole receives at least $\lceil l/p \rceil$ letters and some pigeonhole receives at most $\lfloor l/p \rfloor$ letters.

5.1. Proposition. Given a simple graph G with n vertices and an initial independent set S. At each stage of procedure 3.1, if there are l vertices outside S and the maximum degree among the vertices inside S is less than $\lceil l/(n-l) \rceil$, then procedure 3.1 produces a strictly larger independent set.

Proof. By contradiction. Suppose the independent set S is maximal. Then there are no adjoinable vertices and every vertex outside S must have a neighbor inside S. Thus there are at least l edges (letters) with one end vertex outside S and the other end vertex inside S, there being exactly $p = n-l$ vertices inside S (pigeonholes). By the pigeonhole principle, some vertex inside S must receive at least $\lceil l/p \rceil$ edges contradicting the hypothesis that the maximum degree among the vertices inside S is less than $\lceil l/p \rceil$. □

5.2. Proposition. Given an independent set S of G, procedure 3.1 always produces a maximal independent set of G.

Proof. Procedure 3.1 terminates only when there are no adjoinable vertices. By definition, the resulting independent set must be maximal. □

5.3. Proposition. Given a simple graph G with n vertices and an initial maximal independent set S. If there are m vertices outside the maximal independent set S and the maximum degree among the vertices inside S is less than $\lceil 2m/(n-m) \rceil$, then there exists a vertex v outside S such that v has exactly one neighbor w inside S and procedure 3.2

produces a maximal independent set different from S and of size greater than or equal to the size of S.

Proof. By contradiction. Note that since S is maximal, there are no adjoinable vertices and every vertex outside S has at least one neighbor inside S. Suppose every vertex outside S has more than one neighbor inside S. Then there are at least $l = 2m$ edges (letters) with one end vertex outside S and the other end vertex inside S, there being exactly $p = n-m$ vertices inside S (pigeonholes). By the pigeonhole principle, some vertex inside S must receive at least $\lceil l/p \rceil$ edges contradicting the hypothesis that the maximum degree among the vertices inside S is less than $\lceil l/p \rceil$. Thus, there exists a vertex v outside S such that v has exactly one neighbor w inside S. Now since procedure 3.2 exchanges v and w, an independent set different from S but of the same size as S is created. Note that in the process some vertices outside the independent set might have become adjoinable. Then, procedure 3.2 applies procedure 3.1 that produces a maximal independent set different from S and of size greater than or equal to the size of S. $\square$

5.4. Proposition. Given a simple graph G with n vertices and maximum vertex degree Δ, the algorithm always finds a maximal independent set of size at least $\lceil n/(\Delta+1) \rceil$.

Proof. Consider any one turn of part I in the algorithm. After t vertices have been adjoined from a total of n, there are $l = n-t$ vertices outside the independent set S and the maximum degree among the vertices inside S is certainly less than or equal to Δ. By proposition 5.1, if Δ is less than $\lceil l/(n-l) \rceil = \lceil (n-t)/(n-(n-t)) \rceil = \lceil (n-t)/t \rceil = \lceil (n/t)-1 \rceil$, then a strictly larger independent set is produced by adjoining a vertex. Hence, as long as t is less than $\lceil n/(\Delta+1) \rceil$, a vertex can still be adjoined and procedure 1 continues. Thus, at least $\lceil n/(\Delta+1) \rceil$ vertices are adjoined, producing an independent set of size at least $\lceil n/(\Delta+1) \rceil$. By propositions 5.1, 5.2 and 5.3, all of the independent sets produced by the algorithm are maximal and of size at least $\lceil n/(\Delta+1) \rceil$. $\square$

5.5. Proposition. A simple graph G with n vertices and maximum vertex degree Δ has a maximal independent set of size at least $\lceil n/(\Delta+1) \rceil$.

Proof. By proposition 5.4, the algorithm finds a maximal independent set of size at least $\lceil n/(\Delta+1) \rceil$. $\square$

5.6. Proposition. Given any positive integers n and Δ such that $0 < \Delta < n$, there exists a graph G with maximum vertex degree Δ and a maximum independent set of size $\lceil n/(\Delta+1) \rceil$. For any such graph the algorithm always finds a maximum independent set.

Proof. Let $n = q(\Delta+1)+r$ with $0 \leq r < \Delta+1$ by Euclid's division lemma. There are two cases.

- *Case 1.* Suppose $r = 0$. Define G to be the graph consisting of q disjoint cliques $Q_1, ..., Q_q$ with $\Delta+1$ vertices each. Then G is a graph with maximum vertex degree Δ. Suppose $S_{maximum}$ is a maximum independent set of G. Then $S_{maximum}$ must contain exactly one vertex from each clique, i.e. the size of $S_{maximum}$ must be at most q. On the other hand, by proposition 5.4, the algorithm finds a maximal independent set S of size at least $\lceil n/(\Delta+1) \rceil = \lceil q(\Delta+1)/(\Delta+1) \rceil = q$. Thus, the size of S and $S_{maximum}$ must be the same, i.e. $\lceil n/(\Delta+1) \rceil$.

- *Case* 2. Suppose r is positive. Define G to be the graph consisting of q disjoint cliques $Q_1, ..., Q_q$ with $\Delta+1$ vertices each and a disjoint clique R with r vertices. Then G is a graph with maximum vertex degree Δ. Suppose $S_{maximum}$ is a maximum independent set of G. Then $S_{maximum}$ must contain exactly one vertex from each clique, i.e. the size of $S_{maximum}$ must be at most $q+1$. On the other hand, by proposition 5.4, the algorithm finds a maximal independent set S of size at least $\lceil n/(\Delta+1) \rceil = \lceil (q(\Delta+1)+r)/(\Delta+1) \rceil = q+\lceil r/(\Delta+1) \rceil = q+1$, using the fact that $\lceil r/(\Delta+1) \rceil = 1$ since $0 < r < \Delta+1$. Thus, the size of S and $S_{maximum}$ must be the same, i.e. $\lceil n/(\Delta+1) \rceil$. □

In 1930, Ramsey [6] proved that given any positive integers k and l, there exists a smallest integer $r(k, l)$ such that every graph with $r(k, l)$ vertices contains either a clique of k vertices or an independent set of l vertices. The determination of the Ramsey numbers $r(k, l)$ is in general a very difficult unsolved problem. A *Ramsey graph* $R(k, l)$ is a graph with $n = r(k, l)-1$ vertices that contains neither a clique of k vertices nor an independent set of l vertices. From the definition of the Ramsey numbers it follows that Ramsey graphs $R(k, l)$ exist for all values of k and l greater than 2. We have an immediate

5.7. Corollary. A Ramsey graph $R(k, l)$ with minimum vertex degree δ, maximum vertex degree Δ and $n = r(k, l)-1$ vertices must satisfy

$$\lceil k\delta/(k-1) \rceil < n < l(\Delta+1).$$

Proof. By definition, the graph $G = R(k, l)$ has no clique of size k and no independent set of size l.

By proposition 5.5,

$$\lceil n/(\Delta+1) \rceil < l \quad \Rightarrow \quad n < l(\Delta+1) \qquad (1)$$

On the other hand, by [19] proposition 5.5,

$$\lceil n/(n-\delta) \rceil < k \quad \Rightarrow \quad n < k(n-\delta)$$

$$\Rightarrow \quad n < kn-k\delta$$

$$\Rightarrow \quad k\delta < kn-n$$

$$\Rightarrow \quad k\delta < n(k-1)$$

$$\Rightarrow \quad \lceil k\delta/(k-1) \rceil < n \qquad (2)$$

By (1) and (2), the corollary follows. □

5.8. Question. For all known examples of graphs, the algorithm finds a maximum independent set. In view of the importance of the **P** versus **NP** question **[3]**, we ask: *does there exist a graph for which this algorithm cannot find a maximum independent set?*

6. Implementation

We demonstrate the algorithm with a C++ program following the style of **[7]**. The demonstration program package **[download]** contains a detailed help file and section 7 gives several examples of input/output files for the program.

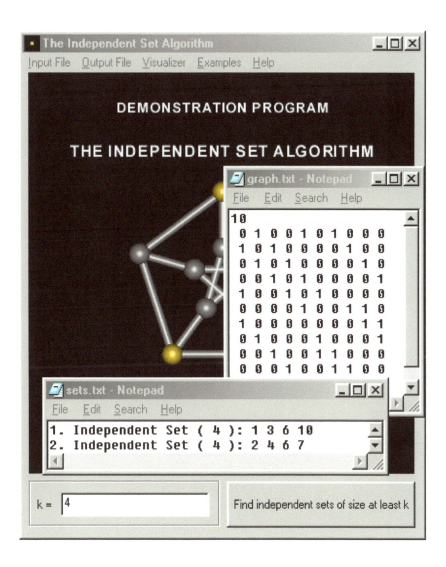

Figure 6.1. Demonstration program for Microsoft Windows **[download]**

17

independent_set.cpp

```cpp
#include <iostream>
#include <fstream>
#include <string>
#include <vector>
using namespace std;

bool removable(vector<int> neighbor, vector<int> cover);
int max_removable(vector<vector<int> > neighbors, vector<int> cover);
vector<int> procedure_1(vector<vector<int> > neighbors, vector<int>
cover);
vector<int> procedure_2(vector<vector<int> > neighbors, vector<int>
cover, int k);
int cover_size(vector<int> cover);
ifstream infile ("graph.txt");
ofstream outfile ("sets.txt");

int main()
{
 //Read Graph
 cout<<"Independent Set Algorithm."<<endl;
 int n, i, j, k, K, p, q, r, s, min, edge, counter=0;
 infile>>n;
 vector< vector<int> > graph;
 for(i=0; i<n; i++)
 {
  vector<int> row;
  for(j=0; j<n; j++)
  {
   infile>>edge;
   row.push_back(edge);
  }
  graph.push_back(row);
 }
 //Find Neighbors
 vector<vector<int> > neighbors;
 for(i=0; i<graph.size(); i++)
 {
  vector<int> neighbor;
  for(j=0; j<graph[i].size(); j++)
  if(graph[i][j]==1) neighbor.push_back(j);
  neighbors.push_back(neighbor);
 }
 cout<<"Graph has n = "<<n<<" vertices."<<endl;
 //Read maximum size of Independent Set wanted
 cout<<"Find an Independent Set of size at least k = ";
 cin>>K; k=n-K;
 //Find Independent Sets
 bool found=false;
 cout<<"Finding Independent Sets..."<<endl;
 min=n+1;
 vector<vector<int> > covers;
 vector<int> allcover;
 for(i=0; i<graph.size(); i++)
 allcover.push_back(1);
 for(i=0; i<allcover.size(); i++)
```

18

```
{
 if(found) break;
 counter++; cout<<counter<<". ";  outfile<<counter<<". ";
 vector<int> cover=allcover;
 cover[i]=0;
 cover=procedure_1(neighbors,cover);
 s=cover_size(cover);
 if(s<min) min=s;
 if(s<=k)
  {
   outfile<<"Independent Set ("<<n-s<<"): ";
   for(j=0; j<cover.size(); j++) if(cover[j]==0) outfile<<j+1<<" ";
   outfile<<endl;
   cout<<"Independent Set Size: "<<n-s<<endl;
   covers.push_back(cover);
   found=true;
   break;
  }
 for(j=0; j<n-k; j++)
 cover=procedure_2(neighbors,cover,j);
 s=cover_size(cover);
 if(s<min) min=s;
 outfile<<"Independent Set ("<<n-s<<"): ";
 for(j=0; j<cover.size(); j++) if(cover[j]==0) outfile<<j+1<<" ";
 outfile<<endl;
 cout<<"Independent Set Size: "<<n-s<<endl;
 covers.push_back(cover);
 if(s<=k){ found=true; break; }
 }
//Pairwise Intersections
 for(p=0; p<covers.size(); p++)
 {
  if(found) break;
  for(q=p+1; q<covers.size(); q++)
  {
   if(found) break;
   counter++; cout<<counter<<". ";  outfile<<counter<<". ";
   vector<int> cover=allcover;
   for(r=0; r<cover.size(); r++)
   if(covers[p][r]==0 && covers[q][r]==0) cover[r]=0;
   cover=procedure_1(neighbors,cover);
   s=cover_size(cover);
   if(s<min) min=s;
   if(s<=k)
    {
     outfile<<"Independent Set ("<<n-s<<"): ";
     for(j=0; j<cover.size(); j++) if(cover[j]==0) outfile<<j+1<<" ";
     outfile<<endl;
     cout<<"Independent Set Size: "<<n-s<<endl;
     found=true;
     break;
    }
   for(j=0; j<k; j++)
   cover=procedure_2(neighbors,cover,j);
   s=cover_size(cover);
   if(s<min) min=s;
   outfile<<"Independent Set ("<<n-s<<"): ";
```

```cpp
    for(j=0; j<cover.size(); j++) if(cover[j]==0) outfile<<j+1<<" ";
    outfile<<endl;
    cout<<"Independent Set Size: "<<n-s<<endl;
    if(s<=k){ found=true; break; }
    }
  }
  if(found) cout<<"Found Independent Set of size at least
"<<K<<"."<<endl;
  else cout<<"Could not find Independent Set of size at least
"<<K<<"."<<endl
  <<"Maximum Independent Set size found is "<<n-min<<"."<<endl;
  cout<<"See sets.txt for results."<<endl;
  system("PAUSE");
  return 0;
}


bool removable(vector<int> neighbor, vector<int> cover)
{
  bool check=true;
  for(int i=0; i<neighbor.size(); i++)
  if(cover[neighbor[i]]==0)
  {
    check=false;
    break;
  }
  return check;
}


int max_removable(vector<vector<int> > neighbors, vector<int> cover)
{
  int r=-1, max=-1;
  for(int i=0; i<cover.size(); i++)
  {
    if(cover[i]==1 && removable(neighbors[i],cover)==true)
    {
      vector<int> temp_cover=cover;
      temp_cover[i]=0;
      int sum=0;
      for(int j=0; j<temp_cover.size(); j++)
      if(temp_cover[j]==1 && removable(neighbors[j], temp_cover)==true)
      sum++;
      if(sum>max)
      {
        max=sum;
        r=i;
      }
    }
  }
  return r;
}


vector<int> procedure_1(vector<vector<int> > neighbors, vector<int>
cover)
{
  vector<int> temp_cover=cover;
  int r=0;
```

```cpp
 while(r!=-1)
 {
  r= max_removable(neighbors,temp_cover);
  if(r!=-1) temp_cover[r]=0;
 }
 return temp_cover;
}


vector<int> procedure_2(vector<vector<int> > neighbors, vector<int>
cover, int k)
{
 int count=0;
 vector<int> temp_cover=cover;
 int i=0;
 for(int i=0; i<temp_cover.size(); i++)
 {
  if(temp_cover[i]==1)
  {
   int sum=0, index;
   for(int j=0; j<neighbors[i].size(); j++)
   if(temp_cover[neighbors[i][j]]==0) {index=j; sum++;}
   if(sum==1 && cover[neighbors[i][index]]==0)
   {
    temp_cover[neighbors[i][index]]=1;
    temp_cover[i]=0;
    temp_cover=procedure_1(neighbors,temp_cover);
    count++;
   }
   if(count>k) break;
  }
 }
 return temp_cover;
}


int cover_size(vector<int> cover)
{
 int count=0;
 for(int i=0; i<cover.size(); i++)
 if(cover[i]==1) count++;
 return count;
}
```

Figure 6.2. C++ program for the independent set algorithm **[download]**

7. Examples

We demonstrate the algorithm by running the program on several famous graphs and two large benchmark graphs with hidden maximum independent sets. In each case, the algorithm finds a maximum independent set in polynomial-time.

7.1. The Tetrahedron [8]. We run the program on the graph of the Tetrahedron with $n = 4$ vertices. The algorithm finds a maximum independent set of size $k = 1$.

graph.txt
```
4
0 1 1 1
1 0 1 1
1 1 0 1
1 1 1 0
```

set.txt
```
Independent Set ( 1 ): 2
```

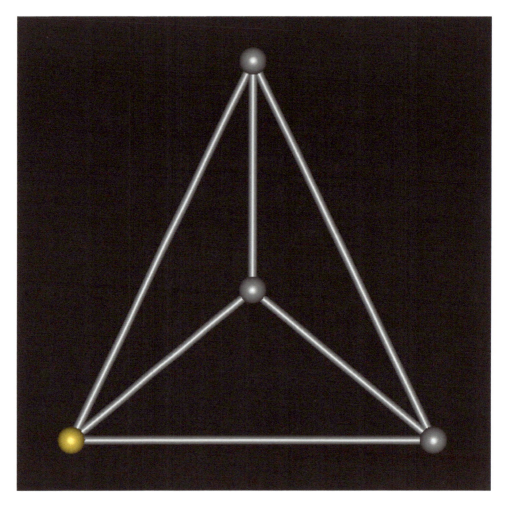

***Figure* 7.1.** *The graph of the Tetrahedron with a maximum independent set*
($n = 4$, $k = 1$).

7.2. The Kuratowski Bipartite Graph $K_{3,3}$ [9]. We run the program on the Kuratowski bipartite graph $K_{3,3}$ with $n = 6$ vertices. The algorithm finds a maximum independent set of size $k = 3$.

graph.txt
```
6
0 0 0 1 1 1
0 0 0 1 1 1
0 0 0 1 1 1
1 1 1 0 0 0
1 1 1 0 0 0
1 1 1 0 0 0
```

set.txt
```
Independent Set ( 3 ): 1 2 3
```

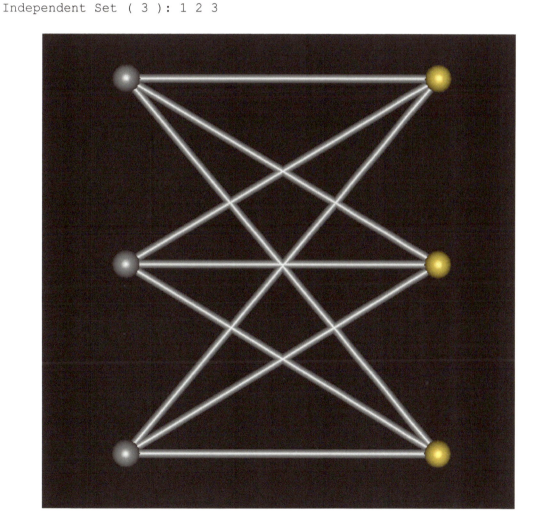

***Figure* 7.2.** *The Kuratowski graph $K_{3,3}$ with a maximum independent set*
($n = 6$, $k = 3$).

7.3. The Octahedron [8]. We run the program on the graph of the Octahedron with $n = 6$ vertices. The algorithm finds a maximum independent set of size $k = 2$.

graph.txt
```
6
0 1 1 0 1 1
1 0 1 1 0 1
1 1 0 1 1 0
0 1 1 0 1 1
1 0 1 1 0 1
1 1 0 1 1 0
```

set.txt
```
Independent Set ( 2 ): 1 4
```

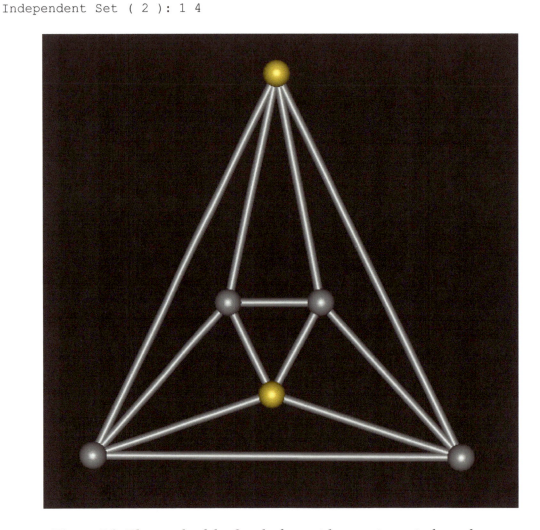

***Figure* 7.3.** *The graph of the Octahedron with a maximum independent set*
(n = 6, k = 2).

7.4. The Bondy-Murty Graph G_1 [4]. We run the program on the Bondy-Murty graph
G_1 with $n = 7$ vertices. The algorithm finds a maximum independent set of size $k = 3$.

graph.txt
```
7
0 1 1 0 1 1 0
1 0 1 1 0 1 0
```

24

```
1 1 0 1 1 0 0
0 1 1 0 0 0 1
1 0 1 0 0 0 1
1 1 0 0 0 0 1
0 0 0 1 1 1 0
```

set.txt
```
Independent Set ( 3 ): 4 5 6
```

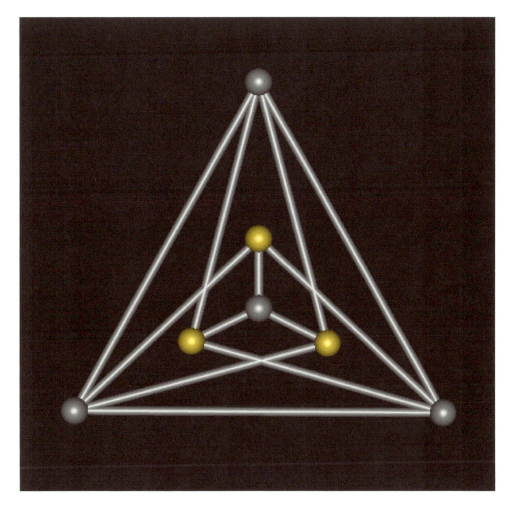

***Figure* 7.4.** *The Bondy-Murty graph G_1 with a maximum independent set*
(n =7, k = 3).

7.5. The Wheel Graph W_8 [4]. We run the program on the Wheel graph W_8 with $n = 8$
vertices. The algorithm finds a maximum independent set of size $k = 3$.

graph.txt
```
8
0 1 0 0 0 0 1 1
1 0 1 0 0 0 0 1
0 1 0 1 0 0 0 1
0 0 1 0 1 0 0 1
0 0 0 1 0 1 0 1
```

25

```
0 0 0 0 1 0 1 1
1 0 0 0 0 1 0 1
1 1 1 1 1 1 1 0
```

set.txt
```
Independent Set ( 3 ): 1 4 6
```

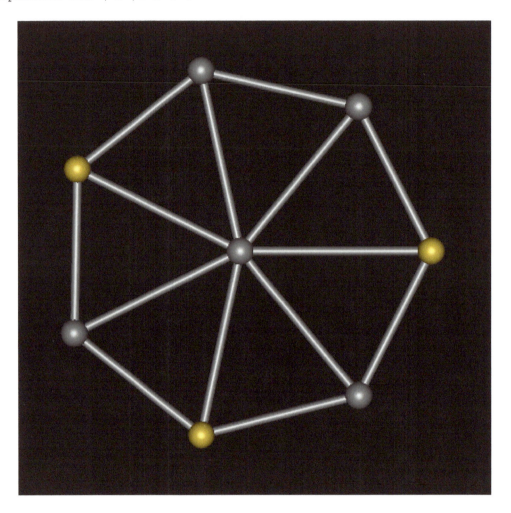

***Figure* 7.5.** *The Wheel graph* W_8 *with a maximum independent set*
($n = 8, k = 3$).

7.6. The Cube [8]. We run the program on the graph of the Cube with $n = 8$ vertices. The algorithm finds a maximum independent set of size $k = 4$.

graph.txt
```
8
0 1 0 1 0 1 0 0
1 0 1 0 0 0 1 0
0 1 0 1 0 0 0 1
1 0 1 0 1 0 0 0
0 0 0 1 0 1 0 1
1 0 0 0 1 0 1 0
```

```
0 1 0 0 0 1 0 1
0 0 1 0 1 0 1 0
```

set.txt
```
Independent Set ( 4 ): 1 3 5 7
```

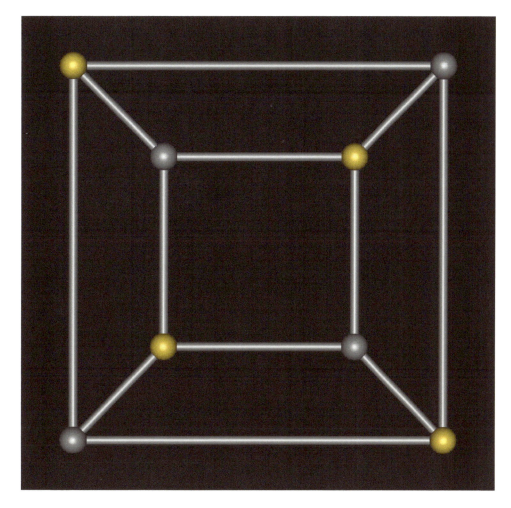

***Figure* 7.6.** *The graph of the Cube with a maximum independent set*
($n = 8$, $k = 4$).

7.7. The Petersen Graph [10]. We run the program on the Petersen graph with $n = 10$
vertices. The algorithm finds a maximum independent set of size $k = 4$.

graph.txt
```
10
0 1 0 0 1 0 1 0 1 0 0 0
1 0 1 0 0 0 0 0 1 0 0
0 1 0 1 0 0 0 0 0 1 0
0 0 1 0 1 0 0 0 0 0 1
1 0 0 1 0 1 0 0 0 0
0 0 0 0 1 0 0 1 1 0
1 0 0 0 0 0 0 0 1 1
0 1 0 0 0 1 0 0 0 1
```

```
0 0 1 0 0 1 1 0 0 0
0 0 0 1 0 0 1 1 0 0
```

set.txt
```
Independent Set ( 4 ):  1  3  6  10
```

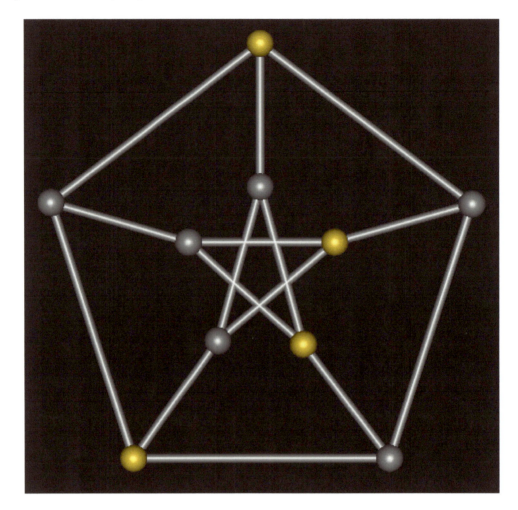

***Figure* 7.7.** *The Petersen graph with a maximum independent set*
($n = 10, k = 4$).

7.8. The Bondy-Murty graph G_2 [4]. We run the program on the Bondy-Murty graph G_2
with $n = 11$ vertices. The algorithm finds a maximum independent set of size $k = 4$.

graph.txt
```
11
0 0 1 1 1 1 0 1 1 1 1
0 0 1 1 1 1 0 1 1 1 1
1 1 0 1 0 0 1 0 0 0 0
1 1 1 0 0 0 1 0 0 0 0
1 1 0 0 0 1 1 0 0 0 0
1 1 0 0 1 0 1 0 0 0 0
0 0 1 1 1 1 0 1 1 1 1
1 1 0 0 0 0 1 0 1 0 0
```

```
1 1 0 0 0 0 0 1 1 0 0 0
1 1 0 0 0 0 0 1 0 0 0 1
1 1 0 0 0 0 0 1 0 0 1 0
```

set.txt
```
Independent Set ( 3 ): 1 2 7
Independent Set ( 3 ): 1 2 7
Independent Set ( 4 ): 3 6 8 11
```

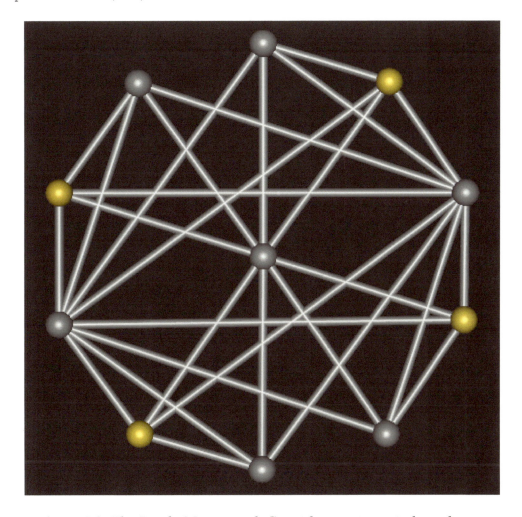

Figure 7.8. *The Bondy-Murty graph* G_2 *with a maximum independent set*
($n = 11, k = 4$).

7.9. The Grötzsch Graph [11]. We run the program on the Grötzsch graph with $n = 11$ vertices. The algorithm finds a maximum independent set of size $k = 5$.

graph.txt
```
11
0 1 1 1 1 1 0 0 0 0 0
1 0 0 0 0 0 1 0 1 0 0
1 0 0 0 0 0 0 1 0 1 0
1 0 0 0 0 0 0 0 1 0 1
1 0 0 0 0 0 1 0 0 1 0
```

29

```
1 0 0 0 0 0 0 1 0 0 1
0 1 0 0 1 0 0 1 0 0 1
0 0 1 0 0 1 1 0 1 0 0
0 1 0 1 0 0 0 1 0 1 0
0 0 1 0 1 0 0 0 1 0 1
0 0 0 1 0 1 1 0 0 1 0
```

set.txt
```
Independent Set ( 5 ): 2 3 4 5 6
```

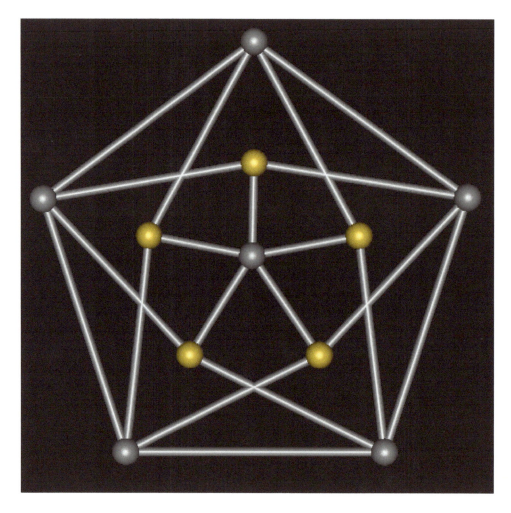

***Figure* 7.9.** *The Grötzsch graph with a maximum independent set*
(n = 11, k = 5).

7.10. The Herschel Graph [12]. We run the program on the Herschel graph with *n* = 11 vertices. The algorithm finds a maximum independent set of size *k* = 6.

graph.txt
```
11
0 1 0 1 1 0 1 0 0 0 0
1 0 1 0 0 0 0 1 0 0 0
0 1 0 1 0 0 0 0 1 0 0
1 0 1 0 0 0 0 0 0 1 0
```

```
1 0 0 0 0 1 0 0 0 1 0
0 0 0 0 1 0 1 0 0 0 1
1 0 0 0 0 1 0 1 0 0 0
0 1 0 0 0 0 1 0 1 0 1
0 0 1 0 0 0 0 1 0 1 0
0 0 0 1 1 0 0 0 1 0 1
0 0 0 0 0 1 0 1 0 1 0
```

set.txt
```
Independent Set ( 5 ): 1 3 6 8 10
Independent Set ( 6 ): 2 4 5 7 9 11
```

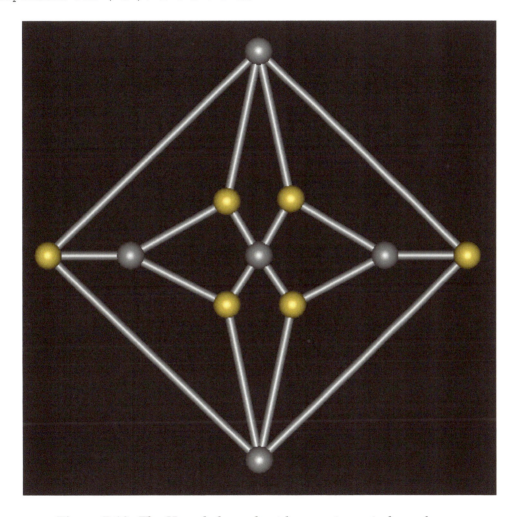

***Figure* 7.10.** *The Herschel graph with a maximum independent set*
($n = 11$, $k = 6$).

7.11. The Icosahedron [8]. We run the program on the graph of the Icosahedron with $n = $ 12 vertices. The algorithm finds a maximum independent set of size $k = 3$.

graph.txt
```
12
 0 1 1 0 0 1 1 1 0 0 0 0
 1 0 1 1 1 1 0 0 0 0 0 0
```

```
1 1 0 1 0 0 0 1 1 0 0 0
0 1 1 0 1 0 0 0 0 1 1 0 0
0 1 0 1 0 1 0 0 0 0 1 1 0
1 1 0 0 1 0 1 0 1 0 0 0 1 0
1 0 0 0 0 1 0 1 0 1 0 0 1 1
1 0 1 0 0 0 1 0 1 0 1 0 0 1
0 0 1 1 0 0 0 1 0 1 0 1 0 1
0 0 0 1 1 0 0 0 1 0 1 0 1 1
0 0 0 0 1 1 1 0 0 1 0 1 0 1
0 0 0 0 0 0 1 1 1 1 1 1 0
```

set.txt
Independent Set (3): 4 8 11

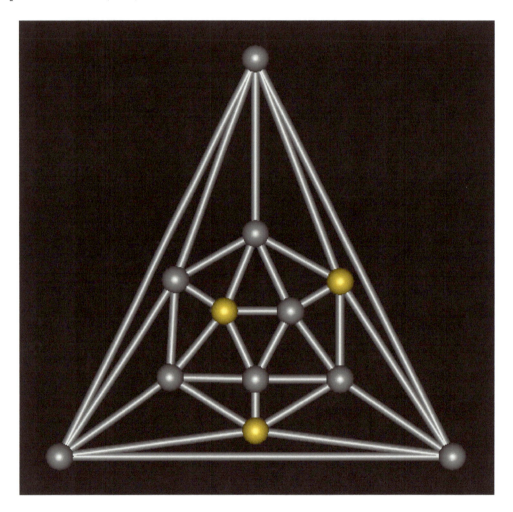

***Figure* 7.11.** *The graph of the Icosahedron with a maximum independent set*
($n = 12$, $k = 3$).

7.12. The Bondy-Murty graph G_3 [4]. We run the program on the Bondy-Murty graph G_3 with $n = 14$ vertices. The algorithm finds a maximum independent set of size $k = 7$.

graph.txt

```
14
0 0 0 1 0 0 0 1 0 0 0 1 0 0
0 0 1 0 0 0 1 0 0 0 1 0 0 0
0 1 0 1 0 0 0 0 0 0 0 0 0 1
1 0 1 0 1 0 0 0 0 0 0 0 0 0
0 0 0 1 0 1 0 0 0 1 0 0 0 0
0 0 0 0 1 0 1 0 0 0 0 0 1 0
0 1 0 0 0 1 0 1 0 0 0 0 0 0
1 0 0 0 0 0 1 0 1 0 0 0 0 0
0 0 0 0 0 0 0 1 0 1 0 0 0 1
0 0 0 0 1 0 0 0 1 0 1 0 0 0
0 1 0 0 0 0 0 0 0 1 0 1 0 0
1 0 0 0 0 0 0 0 0 0 1 0 1 0
0 0 0 0 0 1 0 0 0 0 0 1 0 1
0 0 1 0 0 0 0 0 1 0 0 0 1 0
```

set.txt

```
Independent Set ( 7 ): 1 3 5 7 9 11 13
```

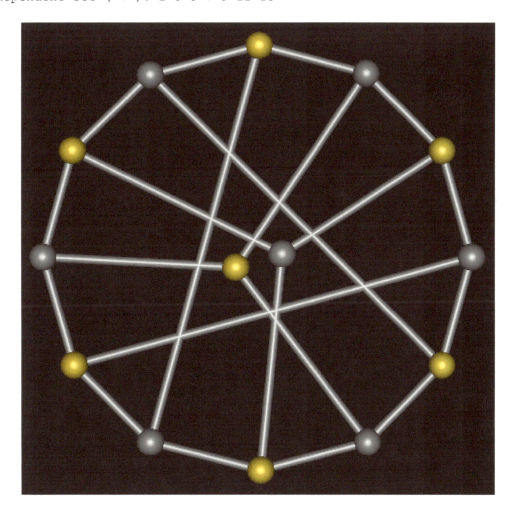

Figure 7.12. *The Bondy-Murty graph G_3 with a maximum independent set ($n = 14$, $k = 7$).*

7.13. The Bondy-Murty graph G_4 [4]. We run the program on the Bondy-Murty graph G_4 with $n = 16$ vertices. The algorithm finds a maximum independent set of size $k = 9$.

graph.txt
```
16
0 0 0 0 0 1 0 0 0 0 0 0 0 1 0 0 0
0 0 0 0 0 0 1 0 0 1 0 0 0 0 0 0 0
0 0 0 0 0 1 0 0 0 0 0 0 1 0 0 0
0 0 0 0 0 0 0 0 0 0 1 0 0 0 0 0
0 0 0 0 0 0 0 1 0 0 0 0 0 0 0 0
1 0 1 0 0 0 0 0 0 0 1 0 0 0 0 0
0 1 0 0 0 0 0 0 1 0 0 0 0 0 0 0
0 0 0 0 1 0 0 0 0 0 0 0 1 0 1 0 0
0 0 0 0 0 0 1 0 0 0 0 0 0 0 1 0
0 1 0 0 0 0 0 0 0 0 0 0 0 0 1 0
0 0 0 1 0 1 0 0 0 0 0 0 0 0 0 0
0 0 0 0 0 0 0 1 0 0 0 0 0 0 0 0
1 0 1 0 0 0 0 0 0 0 0 0 0 0 0 0
0 0 0 0 0 0 0 1 0 0 0 0 0 0 0 0
0 0 0 0 0 0 0 0 1 1 0 0 0 0 0 1
0 0 0 0 0 0 0 0 0 0 0 0 0 0 1 0
```

set.txt
```
Independent Set ( 9 ): 1 3 4 5 9 10 12 14 16
```

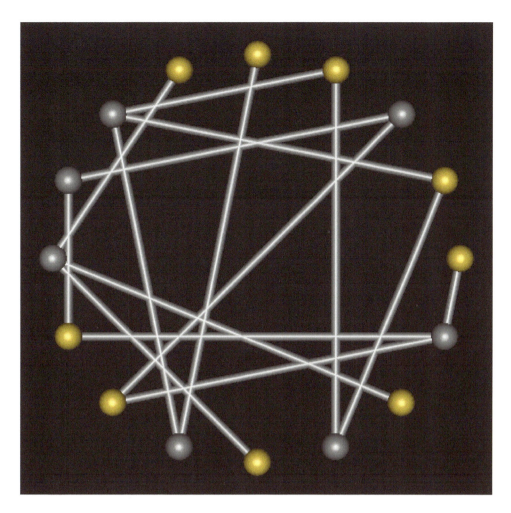

***Figure* 7.13.** *The Bondy-Murty graph G_4 with a maximum independent set*
(n =16, k = 9).

7.14. The Ramsey Graph $R(4,4)$ [6]. We run the program on the Ramsey graph $R(4,4)$
with n = 17 vertices. The algorithm finds a maximum independent set of size k = 3.

graph.txt
```
17
0 1 1 0 1 0 0 0 1 1 0 0 0 1 0 1 1
1 0 1 1 0 1 0 0 0 1 1 0 0 0 1 0 1
1 1 0 1 1 0 1 0 0 0 1 1 0 0 0 1 0
0 1 1 0 1 1 0 1 0 0 0 1 1 0 0 0 1
1 0 1 1 0 1 1 0 1 0 0 0 1 1 0 0 0
0 1 0 1 1 0 1 1 0 1 0 0 0 1 1 0 0
0 0 1 0 1 1 0 1 1 0 1 0 0 0 1 1 0
0 0 0 1 0 1 1 0 1 1 0 1 0 0 0 1 1
1 0 0 0 1 0 1 1 0 1 1 0 1 0 0 0 1
1 1 0 0 0 1 0 1 1 0 1 1 0 1 0 0 0
0 1 1 0 0 0 1 0 1 1 0 1 1 0 1 0 0
0 0 1 1 0 0 0 1 0 1 1 0 1 1 0 1 0
0 0 0 1 1 0 0 0 1 0 1 1 0 1 1 0 1
1 0 0 0 1 1 0 0 0 1 0 1 1 0 1 1 0
0 1 0 0 0 1 1 0 0 0 1 0 1 1 0 1 1
```

```
1 0 1 0 0 0 0 1 1 0 0 0 1 0 1 1 0 1
1 1 0 1 0 0 0 1 1 0 0 0 1 0 1 1 0
```

set.txt
```
Independent Set ( 3 ): 11 14 17
```

***Figure* 7.14.** *The Ramsey graph R(4,4) with a maximum independent set*
($n = 17, k = 3$).

7.15. The Folkman Graph [13]. We run the program on the Folkman graph with $n = 20$ vertices. The algorithm finds a maximum independent set of size $k = 10$.

graph.txt
```
20
0 0 0 0 0 0 0 0 0 0 1 0 1 0 0 0 0 0 1 1
0 0 0 0 0 0 0 0 0 0 0 1 0 1 0 0 0 1 1 0
0 0 0 0 0 0 0 0 0 0 0 0 1 0 1 0 1 1 0 0
0 0 0 0 0 0 0 0 0 0 1 0 0 1 0 1 1 0 0 0
0 0 0 0 0 0 0 0 0 0 0 1 0 0 1 1 0 0 0 1
0 0 0 0 0 0 0 0 0 0 0 1 0 0 1 1 0 0 0 1
0 0 0 0 0 0 0 0 0 0 1 0 0 1 0 1 1 0 0 0
0 0 0 0 0 0 0 0 0 0 0 0 1 0 1 0 1 1 0 0
0 0 0 0 0 0 0 0 0 0 0 1 0 1 0 0 0 1 1 0
```

36

```
0 0 0 0 0 0 0 0 0 0 1 0 1 0 0 0 0 0 0 1 1
1 0 0 1 0 0 1 0 0 1 0 0 0 0 0 0 0 0 0 0 0
0 1 0 0 1 1 0 0 1 0 0 0 0 0 0 0 0 0 0 0 0
1 0 1 0 0 0 0 1 0 1 0 0 0 0 0 0 0 0 0 0 0
0 1 0 1 0 0 1 0 1 0 0 0 0 0 0 0 0 0 0 0 0
0 0 1 0 1 1 0 1 0 0 0 0 0 0 0 0 0 0 0 0 0
0 0 0 1 1 1 1 0 0 0 0 0 0 0 0 0 0 0 0 0 0
0 0 1 1 0 0 1 1 0 0 0 0 0 0 0 0 0 0 0 0 0
0 1 1 0 0 0 0 1 1 0 0 0 0 0 0 0 0 0 0 0 0
1 1 0 0 0 0 0 0 1 1 0 0 0 0 0 0 0 0 0 0 0
1 0 0 0 1 1 0 0 0 1 0 0 0 0 0 0 0 0 0 0 0
```

set.txt
Independent Set (10): 1 2 3 4 5 6 7 8 9 10

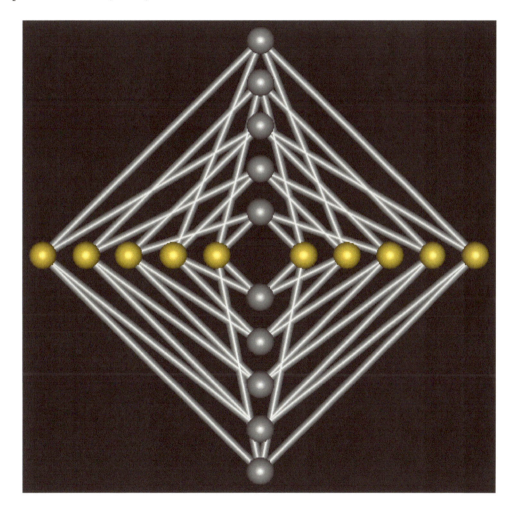

***Figure* 7.15.** *The Folkman graph with a maximum independent set*
($n = 20$, $k = 10$).

7.16. The Dodecahedron [8]. We run the program on the graph of the Dodecahedron with $n = 20$ vertices. The algorithm finds a maximum independent set of size $k = 8$.

graph.txt

```
20
0 1 0 0 1 0 0 0 0 0 0 0 0 0 1 0 0 0 0 0
1 0 1 0 0 0 0 0 0 0 0 1 0 0 0 0 0 0 0 0
0 1 0 1 0 0 0 0 0 1 0 0 0 0 0 0 0 0 0 0
0 0 1 0 1 0 0 1 0 0 0 0 0 0 0 0 0 0 0 0
1 0 0 1 0 1 0 0 0 0 0 0 0 0 0 0 0 0 0 0
0 0 0 0 1 0 1 0 0 0 0 0 0 0 1 0 0 0 0 0
0 0 0 0 0 1 0 1 0 0 0 0 0 0 0 1 0 0 0 0
0 0 0 1 0 0 1 0 1 0 0 0 0 0 0 0 0 0 0 0
0 0 0 0 0 0 0 1 0 1 0 0 0 0 0 0 1 0 0 0
0 0 1 0 0 0 0 0 1 0 1 0 0 0 0 0 0 0 0 0
0 0 0 0 0 0 0 0 0 1 0 1 0 0 0 0 0 1 0 0
0 1 0 0 0 0 0 0 0 0 1 0 1 0 0 0 0 0 0 0
0 0 0 0 0 0 0 0 0 0 0 1 0 1 0 0 0 0 0 1
1 0 0 0 0 0 0 0 0 0 0 0 1 0 1 0 0 0 0 0
0 0 0 0 0 1 0 0 0 0 0 0 0 1 0 1 0 0 0 0
0 0 0 0 0 0 0 0 0 0 0 0 0 0 1 0 1 0 0 1
0 0 0 0 0 0 1 0 0 0 0 0 0 0 0 1 0 1 0 0
0 0 0 0 0 0 0 0 1 0 0 0 0 0 0 0 1 0 1 0
0 0 0 0 0 0 0 0 0 0 1 0 0 0 0 0 0 1 0 1
0 0 0 0 0 0 0 0 0 0 0 0 1 0 0 1 0 0 1 0
```

set.txt

```
Independent Set ( 8 ): 1 3 6 8 11 13 16 18
```

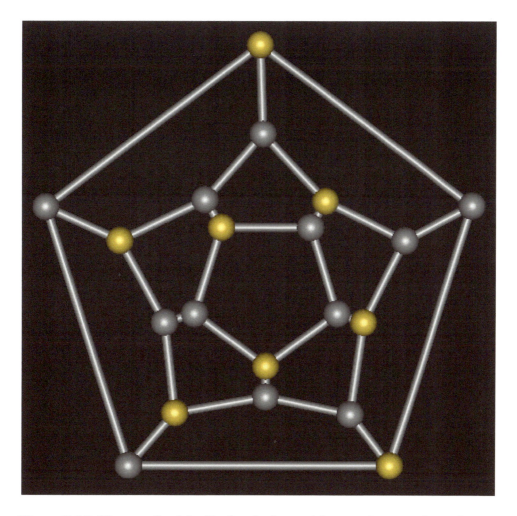

Figure 7.16. *The graph of the Dodecahedron with a maximum independent set*
(n = 20, k = 8).

7.17. The Tutte-Coxeter Graph [14]. We run the program on the Tutte-Coxeter graph
with *n* = 30 vertices. The algorithm finds a maximum independent set of size *k* = 15.

```
graph.txt
30
0 1 0 0 0 0 0 0 0 0 0 0 0 0 0 0 0 0 0 0 0 1 0 0 0 0 0 0 0 1
1 0 1 0 0 0 0 0 1 0 0 0 0 0 0 0 0 0 0 0 0 0 0 0 0 0 0 0 0 0
0 1 0 1 0 0 0 0 0 0 0 0 0 0 0 0 0 0 0 0 0 0 1 0 0 0 0 0 0 0
0 0 1 0 1 0 0 0 0 0 0 0 1 0 0 0 0 0 0 0 0 0 0 0 0 0 0 0 0 0
0 0 0 1 0 1 0 0 0 0 0 0 0 0 0 0 1 0 0 0 0 0 0 0 0 0 0 0 0 0
0 0 0 0 1 0 1 0 0 0 0 0 0 0 0 0 0 0 0 1 0 0 0 0 0 0 0 0 0 0
0 0 0 0 0 1 0 1 0 0 0 0 0 0 0 0 0 0 0 0 0 0 0 0 0 1 0 0 0 0
0 0 0 0 0 0 1 0 1 0 0 0 0 0 1 0 0 0 0 0 0 0 0 0 0 0 0 0 0 0
0 1 0 0 0 0 0 1 0 1 0 0 0 0 0 0 0 0 0 0 0 0 0 0 0 0 0 0 0 0
0 0 0 0 0 0 0 0 1 0 1 0 0 0 0 0 0 1 0 0 0 0 0 0 0 0 0 0 0 0
0 0 0 0 0 0 0 0 0 1 0 1 0 0 0 0 0 0 0 0 0 1 0 0 0 0 0 0 0 0
0 0 0 0 0 0 0 0 0 0 1 0 1 0 0 0 0 0 0 0 0 0 0 0 0 0 0 0 1 0
0 0 0 1 0 0 0 0 0 0 0 1 0 1 0 0 0 0 0 0 0 0 0 0 0 0 0 0 0 0
0 0 0 0 0 0 0 0 0 0 0 0 1 0 1 0 0 0 0 0 1 0 0 0 0 0 0 0 0 0
0 0 0 0 0 0 0 1 0 0 0 0 0 1 0 1 0 0 0 0 0 0 0 0 0 0 0 0 0 0
```

39

```
0 0 0 0 0 0 0 0 0 0 0 0 0 0 1 0 1 0 0 0 0 0 0 0 1 0 0 0 0 0
0 0 0 0 0 0 0 0 0 0 0 0 0 0 1 0 1 0 0 0 0 0 0 0 0 0 0 0 0 1
0 0 0 0 1 0 0 0 0 0 0 0 0 0 1 0 1 0 0 0 0 0 0 0 0 0 0 0 0 0
0 0 0 0 0 0 0 0 1 0 0 0 0 0 1 0 1 0 0 0 0 0 0 0 0 0 0 0 0 0
0 0 0 0 0 0 0 0 0 0 0 0 0 0 0 1 0 1 0 0 0 0 1 0 0 0 0 0 0 0
0 0 0 0 0 0 0 0 0 0 0 1 0 0 0 0 1 0 1 0 0 0 0 0 0 0 0 0 0 0
1 0 0 0 0 0 0 0 0 0 0 0 0 0 0 0 0 1 0 1 0 0 0 0 0 0 0 0 0 0
0 0 0 0 0 1 0 0 0 0 0 0 0 0 0 0 0 0 1 0 1 0 0 0 0 0 0 0 0 0
0 0 0 0 0 0 0 0 0 0 1 0 0 0 0 0 0 0 0 1 0 1 0 0 0 0 0 0 0 0
0 0 0 0 0 0 0 0 0 0 0 0 0 0 1 0 0 0 0 0 0 1 0 1 0 0 0 0 0 0
0 0 1 0 0 0 0 0 0 0 0 0 0 0 0 0 0 0 0 0 0 0 1 0 1 0 0 0 0 0
0 0 0 0 0 0 0 0 0 0 0 0 0 0 0 0 0 1 0 0 0 0 0 1 0 1 0 0 0 0
0 0 0 0 0 0 1 0 0 0 0 0 0 0 0 0 0 0 0 0 0 0 0 0 1 0 1 0 0 0
0 0 0 0 0 0 0 0 0 0 0 1 0 0 0 0 0 0 0 0 0 0 0 0 0 1 0 1 0 1
1 0 0 0 0 0 0 0 0 0 0 0 0 0 0 0 1 0 0 0 0 0 0 0 0 0 0 0 1 0
```

set.txt
```
Independent Set ( 15 ):  1 3 5 7 9 11 13 15 17 19 21 23 25 27 29
```

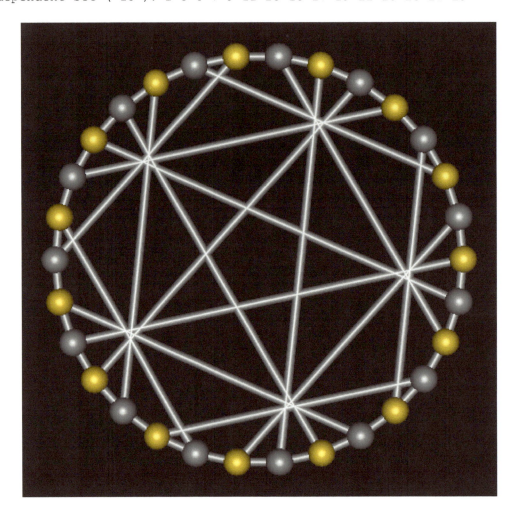

***Figure* 7.17.** *The Tutte-Coxeter graph with a maximum independent set*
(*n* = 30, *k* = 15).

7.18. The Thomassen Graph [15]. We run the program on the Thomassen graph with n = 34 vertices. The algorithm finds a maximum independent set of size $k = 14$.

graph.txt

```
34
0 0 1 1 0 1 0 0 0 0 0 0 0 0 0 0 0 0 0 0 0 0 0 0 0 0 0 0 0 0 0 0 0 0
0 0 0 1 1 0 0 0 0 0 0 0 0 1 0 0 0 0 0 0 0 0 0 0 0 0 0 0 0 0 0 0 0 0
1 0 0 0 1 0 0 0 0 0 0 0 0 0 1 0 0 0 0 0 0 0 0 0 0 0 0 0 0 0 0 0 0 0
1 1 0 0 0 0 0 0 0 0 0 0 0 0 1 0 0 0 0 0 0 0 0 0 0 0 0 0 0 0 0 0 0 0
0 1 1 0 0 0 0 0 0 0 0 0 0 0 0 1 0 0 0 0 0 0 0 0 0 0 0 0 0 0 0 0 0 0
1 0 0 0 0 0 0 1 1 0 0 0 0 0 0 0 0 0 0 0 0 0 0 0 0 0 0 0 0 0 0 0 0 0
0 0 0 0 0 0 0 0 1 1 0 0 1 0 0 0 0 0 0 0 0 0 0 0 0 0 0 0 0 0 0 0 0 0
0 0 0 0 0 1 0 0 0 1 0 1 0 0 0 0 0 0 0 0 0 0 0 0 0 0 0 0 0 0 0 0 0 0
0 0 0 0 0 1 1 0 0 0 1 0 0 0 0 0 0 0 0 0 0 0 0 0 0 0 0 0 0 0 0 0 0 0
0 0 0 0 0 0 1 1 0 0 0 0 0 0 0 0 0 1 0 0 0 0 0 0 0 0 0 0 0 0 0 0 0 0
0 0 0 0 0 0 0 0 1 0 0 1 0 0 0 0 0 0 0 1 0 0 0 0 0 0 0 0 0 0 0 0 0 0
0 0 0 0 0 0 0 1 0 0 1 0 1 0 0 0 0 0 0 0 0 0 0 0 0 0 0 0 0 0 0 0 0 0
0 0 0 0 0 0 1 0 0 0 0 1 0 1 0 0 0 0 0 0 0 0 0 0 0 0 0 0 0 0 0 0 0 0
0 1 0 0 0 0 0 0 0 0 0 0 1 0 1 0 0 0 0 0 0 0 0 0 0 0 0 0 0 0 0 0 0 0
0 0 1 0 0 0 0 0 0 0 0 0 0 1 0 1 0 0 0 0 0 0 0 0 0 0 0 0 0 0 0 0 0 0
0 0 0 1 0 0 0 0 0 0 0 0 0 0 1 0 1 0 0 0 0 0 0 0 0 0 0 0 0 0 0 0 0 0
0 0 0 0 1 0 0 0 0 0 0 0 0 0 0 1 0 1 0 0 0 0 0 1 0 0 0 0 0 0 0 0 0 0
0 0 0 0 0 0 0 0 0 0 0 0 0 0 0 0 1 0 1 0 0 0 0 0 0 1 0 0 0 0 0 0 0 0
0 0 0 0 0 0 0 0 0 0 0 0 0 0 0 0 0 1 0 1 0 0 0 0 0 0 1 0 0 0 0 0 0 0
0 0 0 0 0 0 0 0 0 0 0 0 0 0 0 0 0 0 1 0 1 0 0 0 0 0 0 1 0 0 0 0 0 0
0 0 0 0 0 0 0 0 0 0 0 0 0 0 0 0 0 0 0 1 0 1 0 0 0 0 0 0 0 0 0 0 0 1
0 0 0 0 0 0 0 0 0 0 0 0 0 0 0 0 0 0 0 0 1 0 1 0 0 0 0 0 0 0 0 0 1 0
0 0 0 0 0 0 0 0 0 0 0 0 0 0 0 0 0 0 0 0 0 1 0 1 0 0 0 0 0 0 1 0 0 0
0 0 0 0 0 0 0 0 0 1 1 0 0 0 0 0 0 0 0 0 0 0 1 0 0 0 0 0 0 1 0 0 0 0
0 0 0 0 0 0 0 0 0 0 0 0 0 0 0 0 0 0 0 0 0 0 0 0 0 1 1 0 1 0 0 0 0 0
0 0 0 0 0 0 0 0 0 0 0 0 0 0 0 0 1 0 0 0 0 0 0 0 0 0 1 1 0 1 0 0 0 0
0 0 0 0 0 0 0 0 0 0 0 0 0 0 0 0 0 1 0 0 0 0 0 0 1 0 0 0 1 0 0 0 0 0
0 0 0 0 0 0 0 0 0 0 0 0 0 0 0 0 0 0 1 0 0 0 0 0 0 1 1 0 0 0 0 0 0 0
0 0 0 0 0 0 0 0 0 0 0 0 0 0 0 0 0 0 0 1 0 0 0 0 0 0 1 1 0 0 0 0 0 0
0 0 0 0 0 0 0 0 0 0 0 0 0 0 0 0 0 0 0 0 0 1 0 0 0 0 0 0 1 1 0 0 0 0
0 0 0 0 0 0 0 0 0 0 0 0 0 0 0 0 0 0 0 0 0 0 1 0 0 0 0 0 0 0 1 1 0 0
0 0 0 0 0 0 0 0 0 0 0 0 0 0 0 0 0 0 0 0 0 0 0 1 0 0 0 0 1 0 0 0 1 0
0 0 0 0 0 0 0 0 0 0 0 0 0 0 0 0 0 0 0 0 0 1 0 0 0 0 0 1 1 0 0 0 0 0
0 0 0 0 0 0 0 0 0 0 0 0 0 0 0 0 0 0 0 1 0 0 0 0 0 0 0 1 1 0 0 0 0 0
```

set.txt

```
Independent Set ( 14 ):  3  4  6  7  12  14  17  19  21  24  25  29  32  33
```

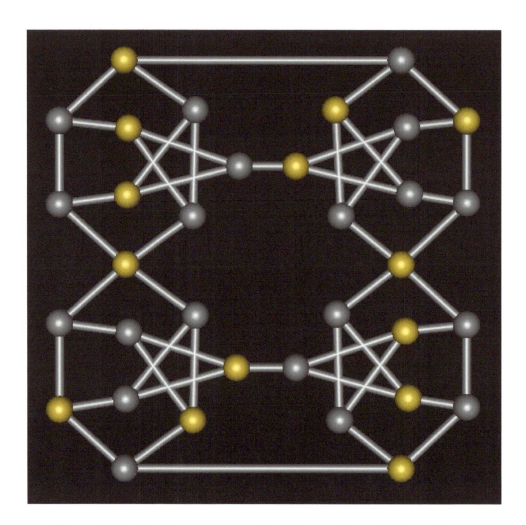

Figure 7.18. *The Thomassen graph with a maximum independent set*
($n = 34, k = 14$).

7.19. The Berge Graph [16]. This is the first benchmark graph with $n = 60$ vertices, following a construction due to Claude Berge. Let G denote the graph of the Dodecahedron and let $H = K_3$ denote the graph of the Triangle i.e. the clique on three vertices. The *Berge graph* $G \times H$ is defined as the graph whose set of vertices is $V(G) \times V(H)$ with an edge connecting vertex (u_1, v_1) with vertex (u_2, v_2) if and only if either $u_1 = u_2$ and $\{v_1, v_2\}$ is an edge in H or $v_1 = v_2$ and $\{u_1, u_2\}$ is an edge in G. It is known that the vertices of the Dodecahedron can be properly coloured with three colours. As a consequence, the Berge graph should have an independent set with at least twenty vertices. Indeed, the algorithm finds a maximum independent set of size $k = 20$.

graph.txt
[download]

set.txt
Independent Set (20): 2 6 7 11 15 16 20 22 26 30 32 34 38 40 45 47 51
52 57 58

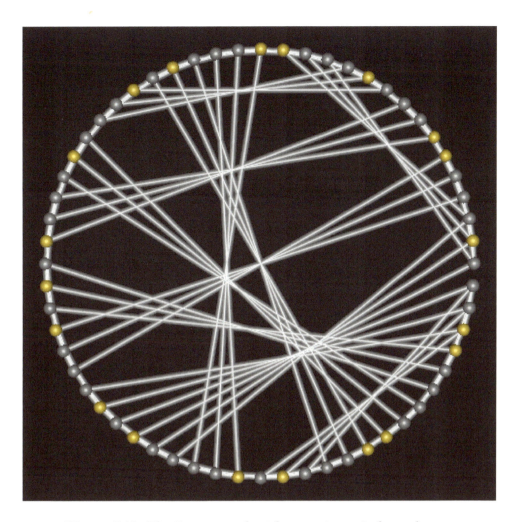

Figure 7.19. *The Berge graph with a maximum independent set*
($n = 60$, $k = 20$).

7.20. The Witzel Graph [17]. This is the second benchmark graph with $n = 450$ vertices, following a construction due to Klaus D. Witzel. Take thirty disjoint cliques on fifteen vertices and connect random pairs of cliques by random edges. Shuffle the labels of the vertices well so that the original cliques are hidden. Provided this is done carefully without adding too many extra edges, such a graph should have a maximum independent set with at least 30 vertices (one vertex from each original clique). Moreover, the maximum independent set is well and truly hidden. Indeed, the algorithm finds a maximum independent set of size $k = 30$.

graph.txt
[download]

set.txt
Independent Set (30): 5 19 34 55 74 78 97 120 122 142 159 167 186 206 211 227 242 268 280 298 314 329 336 351 364 384 400 411 426 442

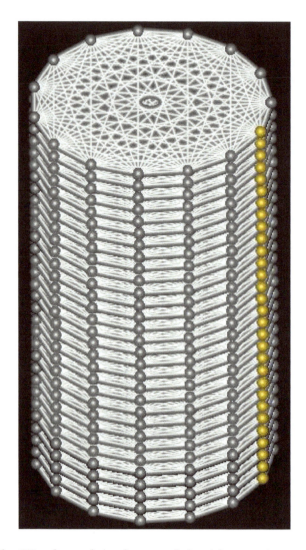

Figure 7.20. *The Witzel graph (scheme only) with a maximum independent set*
(n = 450, k = 30).

8. References

[1] R.M. Karp, *Reducibility among combinatorial problems*, Complexity of Computer Computations, Plenum Press, 1972.

[2] R. Frucht, *Graphs of degree three with a given abstract group*, Canad. J. Math., 1949.

[3] Stephen Cook, *The* **P** *versus* **NP** *Problem*, Official Problem Description, Millennium Problems, Clay Mathematics Institute, 2000.

[4] J.A. Bondy and U.S.R. Murty, *Graph Theory with Applications*, Elsevier Science Publishing Co., Inc, 1976.

[5] Euclid, *Elements*, circa 300 B.C.

[6] F.P. Ramsey, *On a problem of formal logic*, Proc. London Math. Soc., 1930.

[7] Stanley Lippman, *Essential C++*, Addison-Wesley, 2000.

[8] Plato, *Timaeaus*, circa 350 B.C.

[9] K. Kuratowski, *Sur le problème des courbes gauches en topologie*, Fund. Math., 1930.

[10] J. Petersen, *Die Theorie der regulären Graphen*, Acta Math., 1891.

[11] H. Grötzsch, *Ein Dreifarbensatz für dreikreisfreie Netz auf der Kugel*, Z. Martin-Luther-Univ., 1958.

[12] A.S. Herschel, *Sir Wm. Hamilton's Icosian Game*, Quart. J. Pure Applied Math., 1862.

[13] J. Folkman, *Regular line-symmetric graphs*, J. Combinatorial Theory, 1967.

[14] H.S.M. Coxeter and W.T. Tutte, *The Chords of the Non-Ruled Quadratic in PG(3,3)*, Canad. J. Math., 1958.

[15] C. Thomassen, *Hypohamiltonian and hypotraceable graphs*, Discrete Math., 1974.

[16] C. Berge, *Graphes et Hypergraphes*, Dunod, 1970.

[17] Klaus D. Witzel, *Personal Communication*, 2006.

[18] Ashay Dharwadker, *The Vertex Cover Algorithm*, **http://www.dharwadker.org/vertex_cover** , 2006.

[19] Ashay Dharwadker, *The Clique Algorithm*, **http://www.dharwadker.org/clique** , 2006.

[20] Ashay Dharwadker, *The Vertex Coloring Algorithm*, **http://www.dharwadker.org/vertex_coloring** , 2006.